Ford's early factories changed
how cars were made.

12 USES FOR ARTIFICIAL INTELLIGENCE IN TRANSPORTATION

Table of Contents

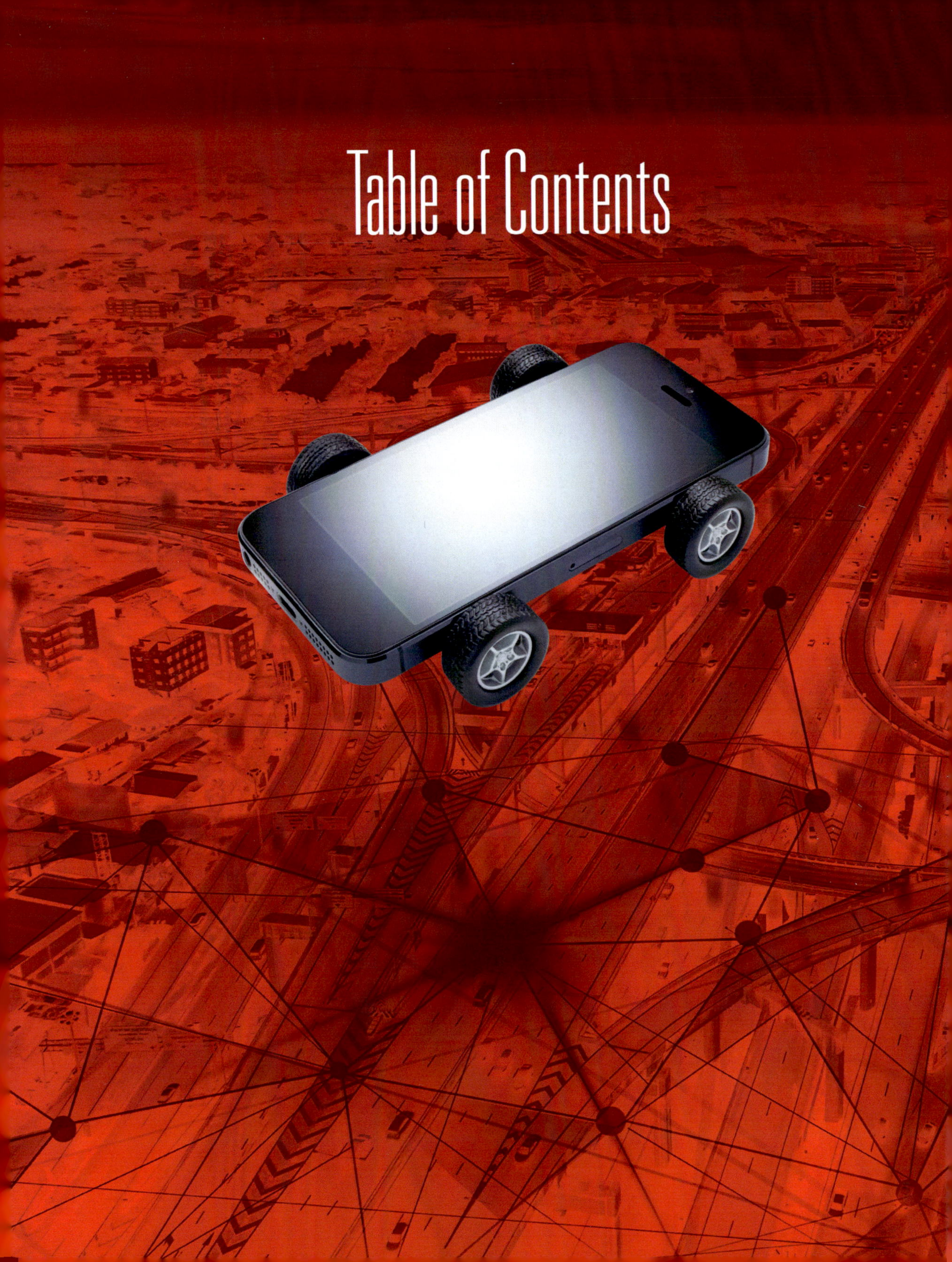

AI Changes the Way Cars *Are Made*

Ford is one of the oldest car companies in the world. It is now ahead of its time. Ford uses artificial intelligence (AI) to look at huge amounts of information. It can check things like car weight, materials, and driving conditions. Then it builds new car models. These cars use less gas. They are safer to drive.

BMW is also using AI. They are working with a high-tech company called NVIDIA. They built a "digital twin" of BMW's factory in Hungary. A digital twin is a computer copy of something. Workers can build cars and fix problems using this computer model. It's all done quickly and with less money thanks to AI.

Tesla uses AI to make cars. First, aluminum is rolled out. Then it's cut out with lasers and stamped. After that, the parts move to the body shop. That's where

AI-powered robotic arms put the car together. It happens very fast. Tesla can build a new car every 13 seconds! Other car companies are copying the way Tesla makes its cars.

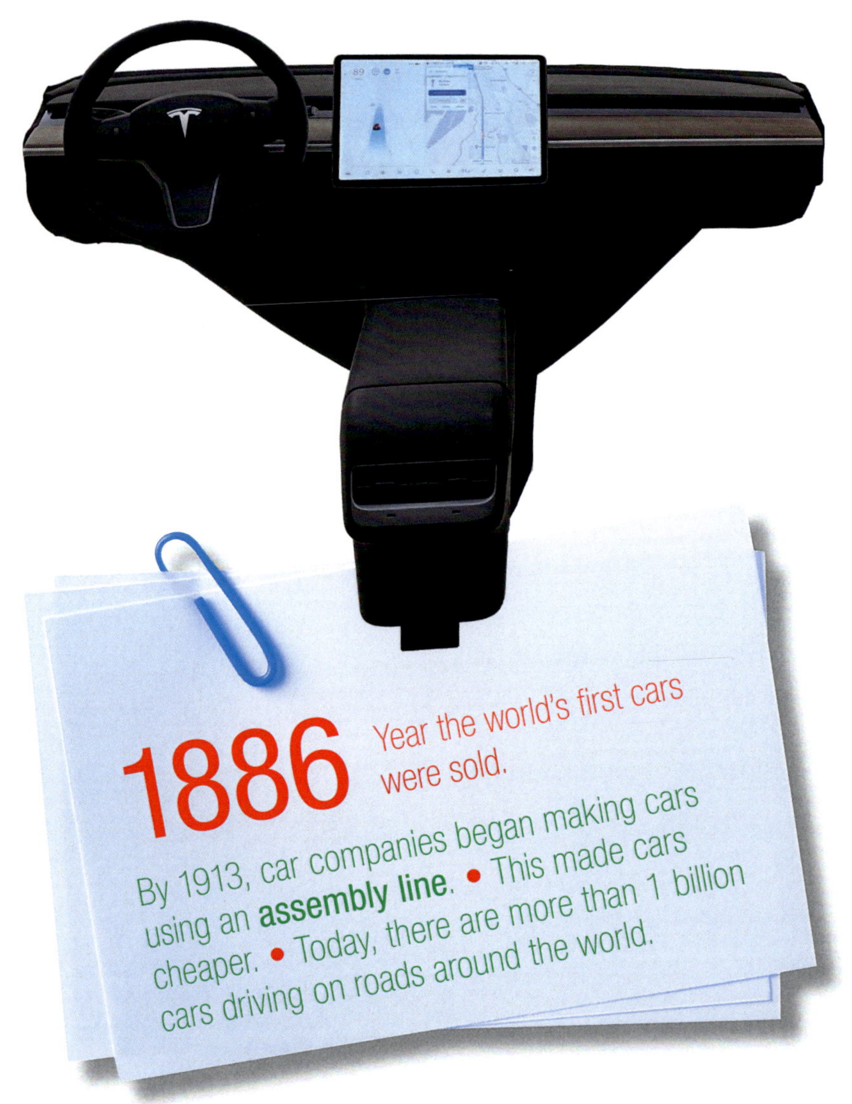

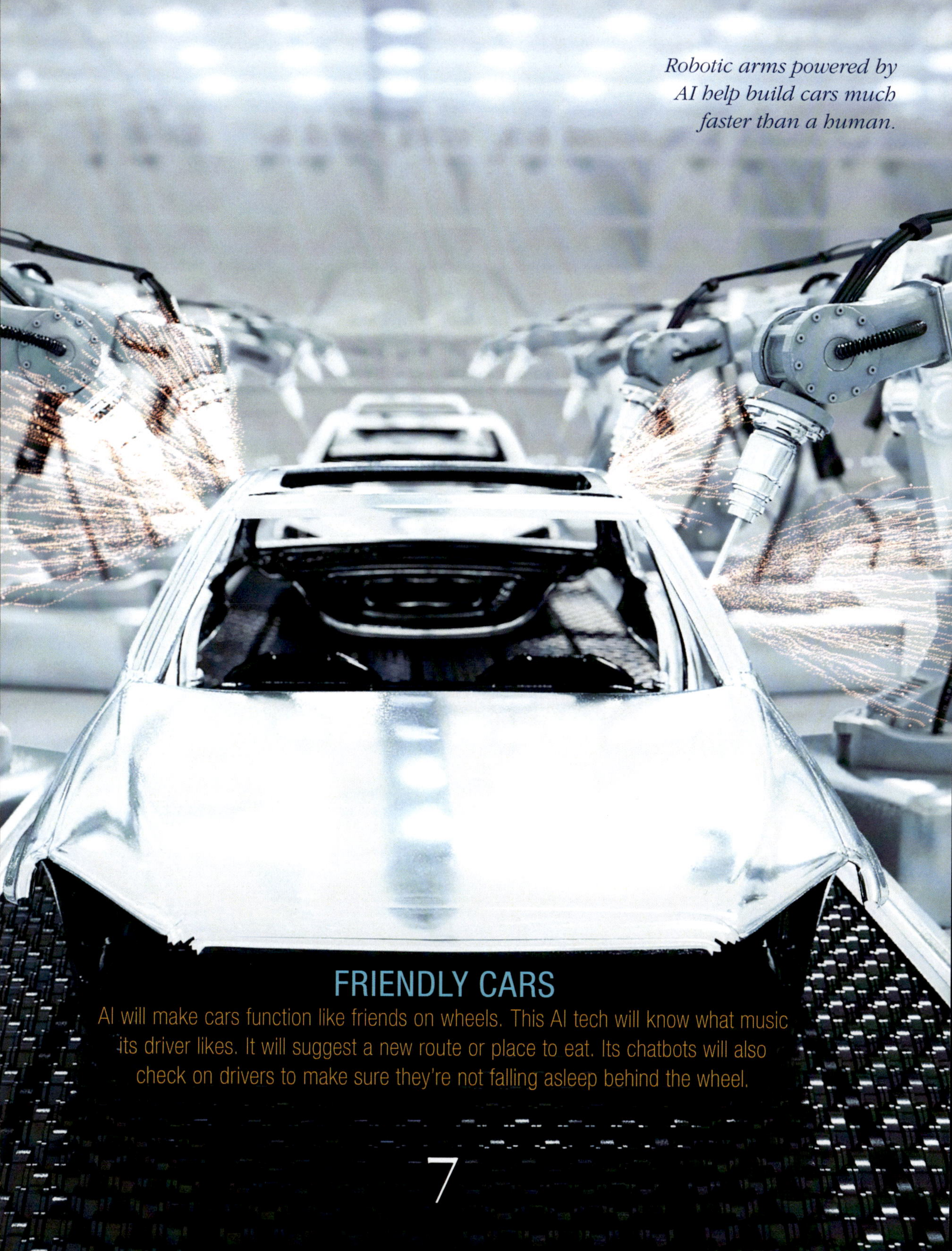

FRIENDLY CARS

AI will make cars function like friends on wheels. This AI tech will know what music its driver likes. It will suggest a new route or place to eat. Its chatbots will also check on drivers to make sure they're not falling asleep behind the wheel.

Using AI Helps Keep *Traffic Moving*

2

Big cities use AI systems to keep cars moving. **Smart cities** have **sensors** and cameras to watch the roads. They send information to traffic centers. They tell traffic lights to stay green. Or they change signs on highways so drivers know if a lane is closed. AI helps people get where they're going. And it helps them get there quickly and safely.

In Pittsburgh, Pennsylvania, smart traffic lights turn red or green based on how many cars are on the road. This helps people drive faster and lessens **greenhouse gases**. In Los Angeles, AI sends emergency vehicles to accidents faster. They clear the roads quickly too.

In the Netherlands, an AI traffic system knows when traffic will get busy. It changes highway speed limits so people can go faster. In Japan, AI looks at weather and traffic. It knows when crashes might happen. Then

9

AI connects cars and traffic systems to reduce jams and crashes.

more police cars are sent out. The speed limit can also be reduced. That way drivers can slow down.

Smart cities are just one way AI helps drivers. Another tool fits in the palm of our hands—our phones! Google Maps and Waze use AI to help drivers plan their trips. They show faster routes. They tell people if there are accidents or closed roads. Apps also tell drivers how long it will take to get places.

AI cameras monitor traffic.

Think About It How do you and your family get to school, work, or your activities? Do you use AI tools to get there faster?

AI Improves Public *Transportation*

Have you ever waited at a bus stop for a long time? Thanks to AI, public transportation is getting better. In Singapore, AI can tell when there's lots of passengers. It can change bus schedules and send more buses when it's busy.

In London, AI can reroute buses if there are problems like construction or accidents. Some cities are testing AI **drones** to watch traffic conditions. These drones show where roads are busy or where more buses are needed.

AI tools, like voice recognition, make it easier for people who are blind to take public transportation. For example, voice alerts announce stops so people know when to get off the bus.

Sensors on a bus or train collect information. Then the AI can tell if there is a loose wire or a problem with the brakes. Workers can repair issues before the bus or train breaks down. Before leaving home, passengers can even use an app called Moovit. This app uses AI to plan the best route and tell people how long it will take to get places. It also lets people pay for their ride.

These forward-thinking solutions improve public transportation. They also make more people want to take a bus or train. Taking the bus is a great way to reduce traffic. It also helps people have a smaller **carbon footprint**, which helps the planet.

MICRO-MOBILITY Micro-mobility transportation are the smaller ways people get around. This includes bikes, scooters, and shared cars. These are easy to use and good for the planet. AI helps make it all work. Companies like SWITCH use AI to keep track of rental bikes and scooters. It makes sure they're in the right place when people need them.

AI tools help make traveling on buses safer and more efficient.
1.7 billion
Number of riders who use Moovit.
Moovit keeps people moving in 3,500 cities across 112 countries. • Moovit has the most information about transportation in the world. • The app helps 7,500 transit operators plan the best routes.

Smart Cars Use AI to *Drive Themselves*

4 Cars that can drive themselves are sometimes called **autonomous** cars. These cars aren't just for the future—they're already here! Self-driving cars use AI to tell if other cars and people are nearby. They also drive at the right speed depending on traffic. And they do it without a human. How does it work? Self-driving cars "see" using cameras and laser-based LiDar (Light Detection And Ranging). Information from these "eyes" is processed using AI. They tell the car where to steer, when to brake, and how fast to go. They can tell if there's a curb or stop sign. They also know if there are traffic lights or construction.

Companies like Kodiak Robotics and Gatik are using self-driving trucks. They move goods—things like clothes or food—long distances. These trucks still have safety drivers on board. Waymo is a self-driving

6

Number of levels for self-driving cars.

Level zero is the first level. A car can't drive itself at this level. • Level five means the car can drive on its own without any help. • In 2024, 60 percent of cars had some driver assistance. • By 2030, up to 10 percent of new cars could be level three or higher.

15

Tesla's AI-powered systems allow for autopilot and self-driving modes.

robotaxi service. It's already driving people around in a few cities, like Phoenix, Arizona, and San Francisco, California. And May Mobility is in several American cities. It uses self-driving Toyota Sienna minivans and no driver.

Car companies like GM, Ford, and Tesla also have self-driving cars. These cars can park and use the emergency brake. They also make sure to stay in their lane. But they still need a human driver behind the wheel

AI Can Call a Ride-Share *Driver*

Ride-share companies like Uber and Lyft are popular. But before people could share cars, people called a taxi. Some people called on a phone. Others stood on the street and waved for a taxi to stop. Now, with AI, getting a ride is much easier. People just use an app on their phone. They can get a car to take them to the subway or anywhere else. They can ride any time of day. People can even have their groceries or food delivered. Apps connect riders to drivers. They also connect restaurants to drivers, and drivers to hungry people.

None of this would be possible without AI. Why? AI does all the scheduling, route planning, and collecting of payments. It collects information and uses it to make travel easy.

AI helps ride-share apps match drivers with riders nearby.

One leader at Uber said AI helps make every ride perfect for each person. AI can learn what riders like and make trips easier and safer. Ride-share apps also help the planet. They can lower traffic and cut pollution by sharing rides. That means fewer people may need cars or big parking lots.

Uber uses AI to improve your riding experience.

ALONG FOR THE RIDE People in some cities can use their phones to call a shuttle service. In Asian countries, a service called Willers makes life less stressful. It has on-demand shuttle buses and self-driving buses. Using their online booking app, riders can even book a trip on a restaurant bus!

New AI tools help taxis plan better routes in cities.

AI Can Plan Better *Delivery Routes*

Delivery companies like Amazon and UPS use AI. It helps them deliver packages faster and more easily. AI helps these companies know how many trucks they need each day. It also helps plan the best routes for drivers to take. This saves time, uses less gas, and helps keep the air clean.

AI also checks road conditions. It helps delivery trucks know about accidents or closed roads. A driver can pick a new route if there's a problem on the road. This way, packages are delivered on time.

But AI doesn't just help with driving. It also powers robots inside big warehouses. These robots help workers easily find, lift, and ship packages. This makes the job easier and speeds up delivery times.

Millions of packages are delivered by Amazon every day.

At Amazon, AI-powered robots are a great help. They find and put away products 75 percent faster than before.

UPS is also making customers happier with AI. They tested a program called Message Response Automation (MeRA). It helped fix 52,000 customer issues. It did it 50 percent faster than a human. That's a big deal. Amazon and UPS both say they will use AI more in the future.

Smart AI systems help get packages to you faster.

AI Is Changing Sea Travel *and Shipping*

7

Traveling the seas is one of the oldest ways to get from place to place. Today, AI is helping make shipping more modern. It collects and studies information so humans can make better decisions. This can save time, fuel, and money.

AI can plan faster and safer shipping routes. It helps run ships, so crews can focus on other important jobs. AI also watches the weather. Crews can change travel plans to avoid hurricanes and other storms.

AI can also track how much fuel a ship is using. AI can come up with ways to save energy. This makes shipping more **sustainable**. It can tell crews when something breaks. AI also knows right away if there's a fire. Small problems can be fixed before they become big ones.

AI guides ships to avoid storms
and find the best routes.

Cameras, heat sensors, and other tools can see what's around the ship, like nearby boats or bridges. AI can watch thousands of hours of video from cameras on the ship to make sure the way is clear. AI also helps analyze cargo on ships. It compares the weight and size to spot anything unsafe. AI can send updates so companies know when their shipments will arrive.

Cargo ships use AI to track shipments and navigate busy ports.

AI Makes Transportation *Safe and Secure*

AI helps police keep people safe. One way is through license plate reader technology. It's a handy way to find stolen or suspicious cars. It's a smart way to catch criminals.

AI is also used for safety on trains, buses, and at airports. It can watch for weapons, strange bags, or anything that looks suspicious. If something seems wrong, AI can tell the police right away.

Facial recognition is another tool powered by AI. It can find a person's face in a crowd. This helps police catch suspects faster. AI also watches oceans and the air with **satellites** and cameras. This keeps travelers safe from bad weather, ships, or other threats.

In emergencies, AI looks at information so rescue teams can get to the scene. This can help save lives.

Car companies like GM, Hyundai, and Volvo work with a company called UVeye. UVeye uses AI to check cars for safety in less than a minute at the factory. Toyota uses a system called Safety Connect. The car can call for help in an accident, even if the driver cannot.

Think About It

Have you ever gone through an airport? Have you seen facial recognition technology? What are some good and bad things about this technology?

Airports use facial recognition to speed up security checks.

AI Is Improving *How We Fly*

9

AI is helping people build better airplanes. These planes can be made on schedule and on budget. AI can also help the environment by making planes that use less fuel.

AI **algorithms** help train pilots. They create flight **simulations**. These are pretend flights that let pilots practice taking off, landing, and handling emergencies. Technology already helps pilots fly planes. In the future, AI may be able to fly planes without a pilot. This could get people around with fewer mistakes or delays.

Airlines and traffic control centers use AI to **predict** the weather. Then they plan the best flight paths. They can also change paths mid-flight if the ride is too bumpy. AI also helps keep planes safe by checking data from sensors. It tells crews when part of a plane needs fixing. And it can do this before anything goes wrong.

AI helps customers too. It shows flight options and
helps people book tickets. It tracks lost bags. AI even
lets people know about gate changes and delays.
One day, AI may even be able to play movies
or shows that each
person likes
to watch.

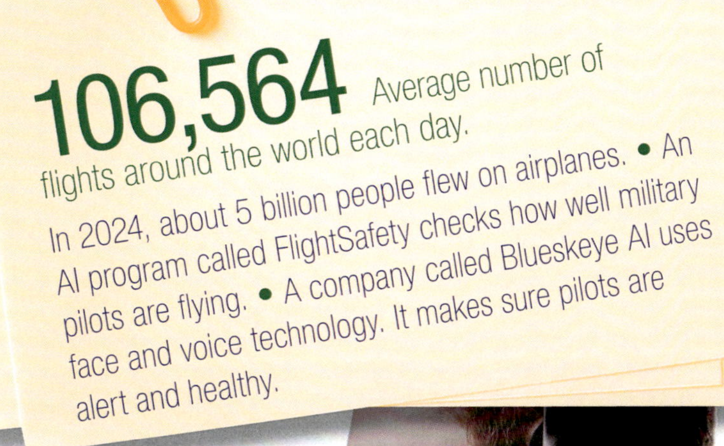

106,564 Average number of flights around the world each day.

In 2024, about 5 billion people flew on airplanes. • An AI program called FlightSafety checks how well military pilots are flying. • A company called Blueskeye AI uses face and voice technology. It makes sure pilots are alert and healthy.

Air traffic controllers use AI to manage busy skies safely.

Pilots use AI-powered simulators to practice flying safely.

AI Makes
Trip Planning
Easy

10

Not too long ago, people had to call a travel agent to book a trip. Today, AI lets people plan trips, book tickets, and get help online. AI can also help people who don't know where they want to go. It learns what people like. It sends them videos on social media. That way people only see what they are interested in.

Websites like Expedia, Kayak, and booking.com use AI. They help people search for the best flight options and prices. This helps them book the best flights. People can also use it to book hotels and rent cars.

A site called Eddy Travels is a personal AI assistant. It uses a chatbot to help people plan their entire trip, from flights to hotels and tours. Trip Advisor is testing an AI virtual voice tour. It works with Amazon and Google assistants. It lets people ask for a voice tour of places to eat, visit, or explore.

30 million

Number of conversations Expedia's virtual agents have with clients.

In 1966, a computer scientist created the first chatbot. This eventually led to virtual agents. • Virtual agents save Expedia about $150 million a year. • The AI has saved the company 8 million hours on phone calls.

AI makes trip planning faster and easier.

Hotels are also getting in on the AI action. Some, like Hyatt, are trying an AI-powered bed. These beds use sensors to track heart rate and breathing. It can change the firmness and temperature of the mattress. That way people can have a softer or a colder bed.

Travelers use AI tools to find places see, eat, and stay.

AI Helps Trains Move *Faster*

Around 1.8 billion people travel by train each year. AI has made their trips better. Many high-speed trains, including **bullet trains**, use AI. Some of these trains can drive themselves. These trains can be found in cities like Paris, Dubai, Tokyo, and Barcelona.

In China, a smart bullet train can make a three-hour trip in 50 minutes. These trains use AI to control the brakes and speed. Now trains run on schedule. Self-driven trains make travel safer. That's because they don't rely on humans who might be tired or make mistakes.

AI watches cameras on board. It can spot problems like theft and lost bags. AI chatbots can help people book tickets or get refunds. This helps keep customers happy.

Bullet trains can travel more than 200 miles (322 kilometers) per hour.

People are excited about hyperloop trains. These are super-fast trains that move through tubes using magnets and air pressure. It's kind of like a vacuum. AI can help design these trains. It can plan routes and decide where stations should go. AI also lets staff know if the trains need fixing.

One day, AI will make it possible for hyperloop trains to drive themselves. They'll be able to stop, speed up, and make decisions. As of 2025, countries like Saudi Arabia, China, and India are working on hyperloop systems. The technology has been tested. But it's not ready for travel just yet. For now, it's still a "pipe" dream!

Hyperloop trains are two to three times faster than bullet trains.

$6 billion
Cost to build a high-speed hyperloop train.

No hyperloop trains exist today. A few companies are working on them, but nobody knows when the first one will be built. • Hyperloop trains may travel up to 700 miles (1,126 km) per hour. • High-speed rail is the fastest kind of train in the world.

Think About It
Why do you think nobody has built a hyperloop train yet? What is stopping the trains from being made?

AI Can Make
Flying Taxis
a Reality

12

AI-powered flying taxis may happen in the near future. These vehicles are also called electric vertical take-off and landing aircraft (eVTOLs). They could start flying in the United Kingdom in 2026. At first, they will be expensive and carry up to five people. It would be like a helicopter. By 2027, some might not need pilots at all. These driverless taxis could deliver medical supplies and mail to far-off places. They might even help chase down criminals.

By 2030, flying taxis could be much more common. Flights would last about 18 minutes. One day, air taxis could carry more people than airlines do today. Some think the flying taxi business could be worth more than $6.6 billion by 2030. Others guess it could be more, making about $33 billion a year.

Before air taxis become common, cities will need to build small airports. New laws and safety rules would need to be made too. People would also need to get used to the idea of flying taxis.

Eventually, AI-powered air taxis could be a faster, safer, and more eco-friendly way to travel. Plus, they could reduce traffic. Experts think air travel will be common by 2045.

MEET THE JETSONS A futuristic cartoon called *The Jetsons* was on TV from 1962–63. It was on TV again in the 1980s. It was the first show in color on ABC. But color wasn't the only thing that people noticed: *The Jetsons* featured flying cars. These cars probably inspired the people creating real flying cars today.

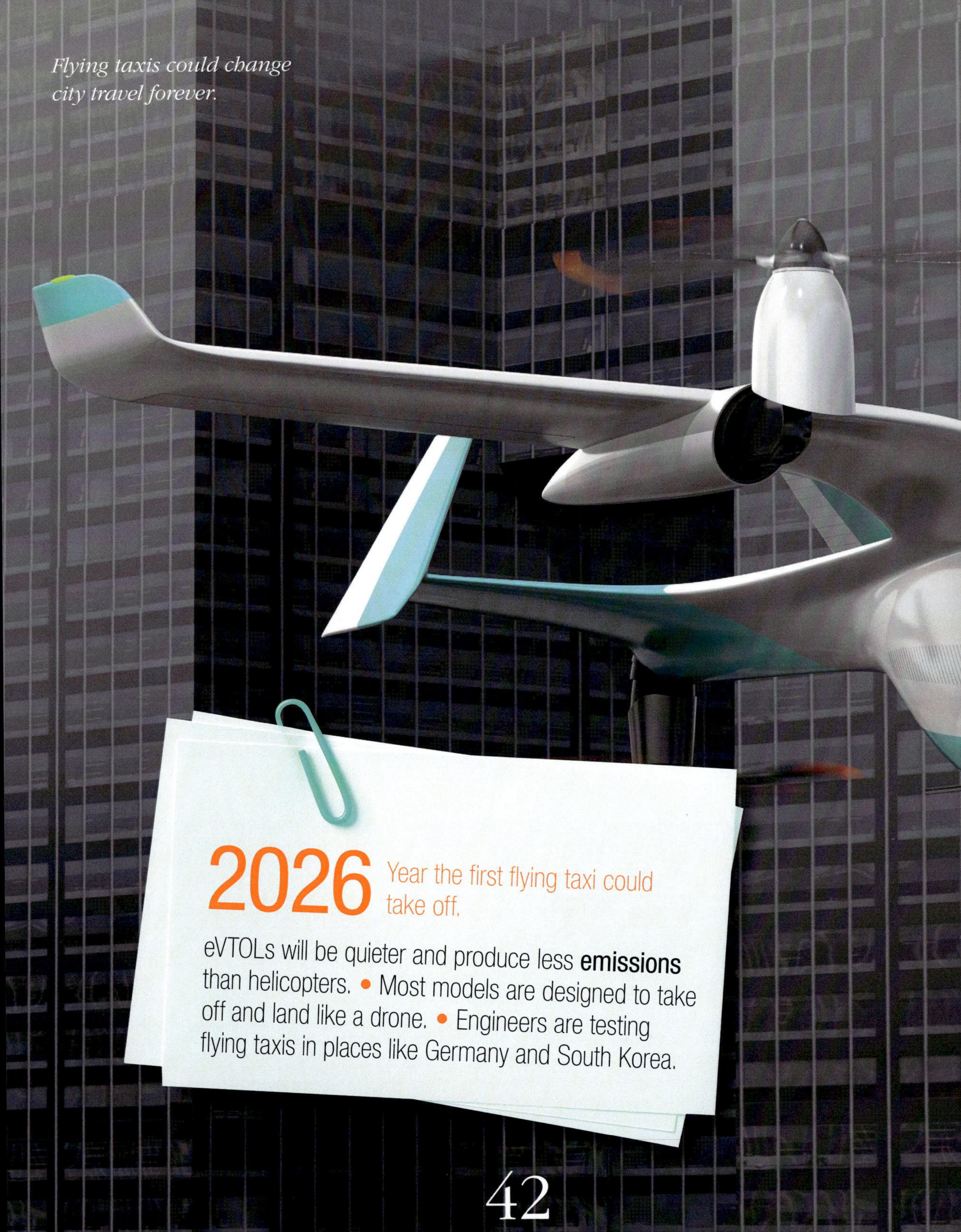
Flying taxis could change city travel forever.

2026
Year the first flying taxi could take off.

eVTOLs will be quieter and produce less emissions than helicopters. • Most models are designed to take off and land like a drone. • Engineers are testing flying taxis in places like Germany and South Korea.

Fact

- In 2025, the digital world will use four times more energy than the whole country of France. The digital world includes things like computers, the internet, and online streaming. As AI helps create self-driving cars and flying taxis, the need for energy will grow. This means we will need cleaner and better ways to use energy.

- The global tech industry produces four percent of the world's greenhouse gas emissions. That's about twice the amount produced by the aviation industry.

Sheet

- At the DARPA Grand Challenge, self-driving cars are tested on different tracks. DARPA is a US government group that creates new technologies for the military. The first self-driving car—a Volkswagen Touareg—drove on the track without a driver in 2005.

- An autonomous robot named *Curiosity* landed on Mars in 2012. It uses an AI program called AEGIS to explore the planet. With laser detection, *Curiosity* looks for signs of life on the red planet. It sends important information to Earth. As of 2025, it is still operating on Mars.

Glossary

algorithm
A set of steps that are followed in order to complete a computer process.

assembly line
An arrangement of workers and machines, each responsible for a step in the process of making a product.

autonomous
Something that can function and complete tasks without being controlled by a human.

bullet train
A high-speed passenger train.

carbon footprint
The amount of carbon dioxide given off by something during a given period.

drone
A vehicle that operates without drivers inside.

emission
Something that is released into the air.

facial recognition
A technology capable of matching a human face from a digital image or a video frame against a database of faces.

greenhouse gas
A gas in Earth's atmosphere that traps heat.

predict
To say something will happen in the future.

satellite
A machine sent into space that moves around a planet, star, or moon.

sensor
A device that detects light, movement, or other information.

simulation
Something that is made to look, feel, or behave like something real.

smart city
An urban area that uses digital technology to collect data and operate services.

sustainable
Capable of continuing or being kept up over time.

For More Information

Books

Cornell, Kari. *Artificial Intelligence and Daily Life.* San Diego: BrightPoint Press, 2025.

Troup, Roxanne. *Robots and AI.* New York: DK Publishing, 2023.

Martin, Emmett. *Self-Driving Cars: Transportation of the Future.* New York: Gareth Stevens Publishing, 2023.

Websites

Autonomous Vehicle
www.britannica.com/technology/autonomous-vehicle

Forecasting with AI

www.timeforkids.com/g56/forecasting-with-ai-g5/?rl=en-900
Humans and AI Working Together
tpt.pbslearningmedia.org/resource/humans-and-ai-working-together-video/crash-course-artificial-intelligence/

About the Author

Erin Silver is an award winning children's author and freelance writer based in Toronto, Canada. She loves learning about new technology and the power of AI. Visit her at ErinSilver.ca.

Index

TOP RANK is published by Black Rabbit Books, P.O. Box 227, Mankato, MN, 56002. • Copyright © 2026 Black Rabbit Books. All rights reserved. No part of this book may be reproduced in any form without written permission from the publisher. • Edited by Ana Brauer • Designed by Danny Nanos • Photographs © Adobe Stock/nerthuz, 38; Alamy Stock Photo/Alasdair Turner, 32–33, PictureLux, 41; Dreamstime/BiancoBlue, 2–3, 9, Cammeraydave, 2, 18–19; Getty Images/onuma Inthapong, 28, Rykoff Collection, 4, Yuichiro Chino, 7; Shutterstock/aerogondo2, 11, Africa Studio, 24, AlinStock, 14, Andrew Angelov, 35, Anna Zhuk, 17, 20, Applepy, 37, Art of pixels, 45, AU USAnakul, 30, 31, ayyse, 24, bbernard, 12–13, Belinda Pretorius, 48, bluestork, 16, Chesky, 42–43, Flystock, 15, Frederic Legrand - COMEO, 21, gofra, 6, GreenOak, 26, Hadrian, 22, 23, heychli, 44, Jolygon, 46–47, Ljupco Smokovski, 10, 44, Lost_in_the_Midwest, 16, Maor_Winetrob, 11, Max Acronym, 31, Mikhail Leonov, 8, Milkovasa, 29, moxumbic, cover, 1, mr. teerapon tiuekhom, 30, NASA images, 45, New Africa, 27, 29, Olga Maksimava, 10, Semachkovsky, 25, skodonnell, 21, Stokkete, 12, Tero Vesalainen, 17, Tony Stock, 28, u3d, 39, vladiwelt, 24, Yeti studio, 34, yuyangc, 37, zeljkodan, 36; Wikimedia Commons/Wolff Olins, in collaboration with Uber's Brand Team, 19 • Printed in the United States of America.

Library of Congress Cataloging-in-Publication Data: Names: Silver, Erin, 1980- author | Title: 12 uses for artificial intelligence in transportation / Erin Silver. | Description: Mankato, MN: Top Rank, an imprint of Black Rabbit Books, [2026] | Series: AI in the world | Includes bibliographical references and index. | Audience: Ages 9–13 | Audience: Grades 4–6 | Identifiers: LCCN 2025021460 (print) | LCCN 2025021461 (ebook) | ISBN 9781645825197 library binding | ISBN 9781645825371 paperback | ISBN 9781645825555 ebook | Subjects: LCSH: Transportation—Technological innovations—Juvenile literature | Artificial intelligence—Transportation—Juvenile literature | Intelligent transportation systems—Juvenile literature | Classification: LCC TA1149 .S56 2026 (print) | LCC TA1149 (ebook) | DDC 625.7/94028563—dc23/eng/20250714 | LC record available at https://lccn.loc.gov/2025021460 | LC ebook record available at https://lccn.loc.gov/2025021461